Playful Animals

CROWS

Ursula Pang

PowerKiDS press

PK Beginners

Crows are birds.

Crows are black.

Crows are loud!

Crows can fly.

Crows live in groups.

Crows lay eggs.

Crows are smart.

Crows can use tools!

Crows remember people they like.

Crows remember people they do not like!

Crows are playful birds!

Published in 2025 by The Rosen Publishing Group, Inc.
2544 Clinton Street, Buffalo, NY 14224

First Edition

Editor: Greg Roza
Book Design: Michael Flynn

Photo Credits: Cover, p. 1 Nicholas Tofanelli/Shutterstock.com; p. 3 Rudmer Zwerver/Shutterstock.com; p. 5 Nancy Bauer/Shutterstock.com; p. 7 Smileynulk/Shutterstock.com; p. 9 Jean Blom/Shutterstock.com; p. 11 Janos Hada/Shutterstock.com; p. 13 Genc Subhan/Shutterstock.com; p. 15 Vladimir Turkenich/Shutterstock.com; p. 17 Brian E Kushner/Shutterstock.com; p. 19 IrinaK/Shutterstock.com; p. 21 Audra Thomson/Shutterstock.com; p. 23 Heidi Bollich/Shutterstock.com.

Library of Congress Cataloging-in-Publication Data

Names: Pang, Ursula, author.
Title: Crows / Ursula Pang.
Description: Buffalo : PowerKids Press, [2025] | Series: Playful animals
Identifiers: LCCN 2024028035 (print) | LCCN 2024028036 (ebook) | ISBN 9781499450712 (library binding) | ISBN 9781499450705 (paperback) | ISBN 9781499450729 (ebook)
Subjects: LCSH: Crows–Juvenile literature.
Classification: LCC QL696.P2367 P36 2025 (print) | LCC QL696.P2367 (ebook) | DDC 598.8/64–dc23/eng/20240621
LC record available at https://lccn.loc.gov/2024028035
LC ebook record available at https://lccn.loc.gov/2024028036

Manufactured in the United States of America

CPSIA Compliance Information: Batch #CWPK25. For further information contact Rosen Publishing at 1-800-237-9932.